Quantum-Dot Laser Behavior

Dr. Fabio Lozano

Kelsiehanson Verlag

DEDICATION

To My Wonderful Readers

CONTENTS

ACKNOWLEDGMENTS

Thank you to everyone who strives to grow and help others grow

1 ABSTRACT

This work explores the bifurcation scenarios of the quantum-dot (QD) semiconductor laser with an external cavity (optical feedback) using a dimensionless model where the dynamics of the ground and the excited states are examined in addition to the wetting layer (WL).It is shown that while chaos was reasoned, mainly, to ground state (gs) dynamics at low injection current as a result of fast interdot relaxation, it is reasoned to WL at high injection due to saturation. Time series for WL, gs, and excited state (es), and the bifurcation diagram are examined at different parameters.

Index Terms— quantum-dot (QD) laser, optical feedback,nonlinear dynamics, chaos control.

2 INTRODUCTION

One of the specific characteristics of semiconductor lasers is their low sensitivity to any external interference as an optical input that is favored in certain scientific applications [1]. In the case of the use of semiconductor lasers in communication and optical networks, expensive optical insulators are required to prevent reflections that may lead to temporal laser instability (coherence collapse) [2].On the other hand, many applications take advantage of the rich dynamics produced by optical feedback, such as chaos key distribution and safe chaos communication[3-6]. Opticalfeedback can be used with some techniques to substantially increase the bandwidth of the directly modulated laser modulation[7]. Delay can cause rich phenomena results in high dimensionality structures ranging from erratic amplitude dropouts (low-frequency fluctuations LFF), multi-stable, sporadic, bursting, and full-blown chaos[8].Semiconductor laser dynamics have been monitored and used to clarify the stability of CW emissions (steady states) or self-pulsation (periodic oscillations) by delay time control feedback [9-16] while delay synchronization and coupled lasers have been investigated by bubbling[17-19].

High dynamical stability was shown by the QD lasers for optical feedback compared to the traditional quantum well (QW) semiconductor laser[20-24]. A QD laser transmitter allowsthe operation without costly optical insulators[25]. Moreover, because of their improved dynamic stability, the road to chaos in the QD laser can be seen more clearly.The efficiency of the QDs improved with an external cavity was coupled with increased oscillations of relaxation damping and miniaturized phase-amplitude [26-29].The model presented here considering boththe QD ground state (gs) and the excited state (es) with occupationsρ_{gs}, ρ_{es}, and the single-state WL population N_{wl}.The presumption in this model is that the carriers are immediately injected into WL, so that they can be captured in QDs. The model neglects the transport of the carrier through the active layer, such an assumption has been usedearlier [30].

The rest of the paper is structured as follows: in Sec. II We presented our QD laser model with an external cavity. Section III iscontaining the results and their discussion. The effect of the linewidth enhancement factor was also studied. Finally, the results of this work were submerged in Section IV.

3 MATERIALS AND METHODS

I. Quantum dot laser model

In the QD laser dynamics, see Fig. 1, carriers are first injected into the QW WL until they are caught in the QD [31, 32]. However, the rate of the capture depends on the degree of occupancy of the QDs due to Pauli blocking. The model presented here is cast from the model in [26]which considers QD with gs only. It is defined by the rate equations of the QD laser subjected to an optical feedback andis consist of four rate equations: Equation for the normalized laser field E in the cavity,two equations for the occupation probabilities ρ_{gs} *and* ρ_{es} *of the gs and the es, respectively, in the QD, and an equation for the number* *of carriers in the WL reservoir N_{wl}. Thesystem is presented by the following rate equations:*

$$\dot{E} = -\frac{1}{2}\gamma_s E - \frac{1}{2}(1+i\alpha)\upsilon g_o(2\rho_{gs}-1)E + \frac{1}{2}\gamma E_{\tau} e^{-i\Theta} \quad \textit{(1.a)}$$

$$\dot{\rho}_{gs} = \gamma_{c_{es}}\rho_{es}(1-\rho_{gs}) - \gamma_d\rho_{gs} - g_o(2\rho_{gs}-1)|E|^2 \quad \textit{(1.b)}$$

$$\dot{\rho}_{es} = \gamma_{c_{wl}}N_{wl}(1-\rho_{es}) - \gamma_d\rho_{es} - \gamma_{c_{es}}\rho_{es}(1-\rho_{gs}) \quad \textit{(1.c)}$$

$$\dot{N}_{wl} = \frac{J}{e} - \gamma_n N_{wl} - 2\gamma_{c_{wl}}N_{wl}(1-\rho_{es}) \quad \textit{(1.d)}$$

Thedot ()shown above on the left-hand side of the system refers to the time derivative. $N_d$is the two-dimensional QDdensity, w_o denoting the frequency of the laser at the lasing threshold, while τ is the external round-trip time(delay time). Since the model is entirely studiedunder $w_o$and τ variation,thefeedback phase $\Theta$is servedas an independent parameter since the change in the length of the external cavity inducesa minor difference in the phase Θ over its full range [0; 2π]while τin theexternal cavity ($\tau = 2L/c$) is hardlyaffectedby these fluctuations.

The phase shift is given by $\Theta = \omega_0 \tau$ *where c is the speed of light.The optical phase* ϕ_τ *and the amplitude of the electrical field* E_τ *are taken at the timedelayed (t-τ). The parameter γis the power of the injected field,* $\upsilon = 2N_d \Lambda / d$ *where* Λ *is the optical confinement factor and d is the thickness of the QD layer.The parameter α is the linewidthenhancementfactor, the gain factor is* $g_o = \sigma v_g$ *where σ is the cross-section of the interaction between carriers and photons in the QDs and* v_g *is the group velocity.Thephoton decay ratesin the cavity,in the WL and the QDare, respectively,* γ_s, γ_n, *and* γ_d. *The capture rate from the WL into an empty excited state is* γ_{cwl}, *while from excited to the gs is* γ_{ces}, *J is the pump current per QD and q is the elementary charge.The terms* $[\gamma_{c_{es}} \rho_{es} (1 - \rho_{gs})]$ *in Eq. (1.c) identifies the rate of exchange of carriers in the QDs from es to gs while the term* $[2\gamma_{c_{wl}} N_{wl} (1 - \rho_{es})]$ *in Eq. (1.d) is that from es into the WL.* $E(t)(= \sqrt{S} e^{-i\Phi(t)})$ *is the electrical fieldand can be represented by the slowly varying normalized complex amplitude of the field Swhich is the photon numberand Φis the phase.Converts the electric field into the photon density S and phase Φ, the system becomes:*

$$S^{\square} = [\upsilon g_o (2\rho_{gs} - 1) - \gamma_s]S + \gamma\sqrt{SS_\tau}\cos(\phi - \phi_\tau)$$

(2.a)

$$\phi^{\square} = -\frac{\alpha}{2}\upsilon g_o (2\rho_{gs} - 1) - \frac{\gamma}{2}\sqrt{S_\tau / S}\sin(\phi - \phi_\tau) \quad (2.b)$$

$$\rho^{\square}_{gs} = \gamma_{c_{es}} \rho_{es} (1 - \rho_{gs}) - \gamma_d \rho_{gs} - g_o (2\rho_{gs} - 1)S \quad (2.c)$$

$$\rho^{\square}_{es} = \gamma_{c_{wl}} N_{wl} (1 - \rho_{es}) - \gamma_d \rho_{es} - \gamma_{c_{es}} \rho_{es} (1 - \rho_{gs})$$

(2.d)

$$N^{\square}_{wl} = \frac{J}{e} - \gamma_n N_{wl} - 2\gamma_{c_{wl}} N_{wl} (1 - \rho_{es}) \quad (2.e)$$

The carrier-light interaction is covered in the photon density S which involves all the longitudinal modes.The factor 2 in Eq. (2.e)accounts for the degeneracy of double spin in the energy levels of the QD and a similar factor (2) is included in the definition of the gainfactor g_o ***in Eq. (2.a). It is useful to rewrite system Eqs. (2) in a dimensionless manner***

where the numerical solution becomes easy. To this end, we introduce the new variables,

$$x = \frac{g_o}{\gamma_d} S \ , \ \Phi \equiv \Phi, \ y = \frac{g_o \upsilon}{\gamma_s}(2\rho_{gs} - 1),$$

$$z \equiv \rho_{es} \ , \ w = \frac{\gamma_{c_{wl}}}{g_o \upsilon} N_{wl} \ ,$$

$$\Gamma = \frac{\gamma_{c_{es}}}{\gamma_s}, \ \Gamma_1 = \frac{g_o \upsilon}{\gamma_s}, \Gamma_2 = \frac{\gamma_d}{\gamma_s},$$

$$\Gamma_3 = \frac{\gamma_{c_{wl}}}{\gamma_s}, \ \Gamma_4 = \frac{\gamma_n}{\gamma_s},$$

$$\delta_o = \frac{J}{g_o \upsilon q}$$ ***and the time scale*** $t' = \gamma_s t$.

The rate equations system,Eq. (2), becomes

$$x' = x(y - 1) + \varepsilon \sqrt{x \, x_\tau} \cos(\phi - \phi_\tau) \quad \textbf{\textit{(3.a)}}$$

$$\Phi' = -\frac{\alpha}{2}y - \frac{\varepsilon}{2}\sqrt{x_\tau/x}\,\sin(\phi - \phi_\tau) \quad \textit{(3.b)}$$

$$y' = \Gamma z(\Gamma_1 - y) - \Gamma_2 y(1 + 2x) - \Gamma_1\Gamma_2 \quad \textit{(3.c)}$$

$$z' = \Gamma_1 w(1 - z) - \Gamma_2 z - \Gamma z(1 - y/\Gamma_1)/2 \quad \textit{(3.d)}$$

$$w' = \Gamma_3\delta_o - \Gamma_4 w - 2\Gamma_3 w(1 - z) \quad \textit{(3.e)}$$

Where $\varepsilon = \gamma/\gamma_s$ *and the assumption here is that the delay time* τ *(in the external cavity) inside the active region is greater than the laser round trip time. The fourth-order Runge-Kutta method is used in numerical simulations, along with Graphics Berkeley Madonna, to analyze the time series created by the chaotic regime. The work includes the examination of the attractors and the resulting laser bifurcation scenario.*

3. Results

The dynamics of QD laser must be covered by examining theoccupations of the continuous higher energy WL,which is considered very high, and the discrete states located at the QDs. Accordingly,the capture dynamics are dominated through thecarrier-photon interaction in the QD-WL.

Table 1 lists the parameters used in the simulation. The speed of the carrier transition from, and to, the excitedstate is determined by the normalized rates Γ and $\Gamma3$ values.Fig. 2 (a) and (b) shows the effect of Γ on the turn-on dynamics of the QD laser without optical feedback where the turn-on delay time decreases with increasing the capture rate from es to an empty gs and the lasing reaches faster, abehavior similar to that in [٢٢, ٢٣].At a low capture rate

(Γ=0.72), a change of the equilibrium point was shown.Low Γ value leads to long turn-on delay time due to a slowdown in the speed at which the charge carriers move from the WL to the es. IncreaseΓ value up to 8.12 is significant in reducing the turn-on delay timewhereas the increase or decrease in the $\Gamma 3$ values does not affect the behavior and this indicates that the es plays the major role in stabilizing the system.This is with the result in ref. [33] that the long QD relaxation time gives less sensitivity to intensity noise.

Fig. 3 shows the bifurcation diagram of photons versus the bias current at fixed feedback strength (0.312).Increasing the bias current results in a chaotic behavior which is then reduced with more increment in the bias to quasi-periodic, double periodic, and then periodic, i.e. the laser becomes less sensitive to optical feedback with increasing bias current. The beginning of this behavior was discovered in [34], while the overall behavior is similar to that seen in bulk lasers by Ohtsubo[35].This can be reasoned to the type of QD layers that construct a QD laser. Although WL has a higher number of carriers-continuum state, the QD layer has a very few number of states. The QD gs was quickly filled and empty due to the rapid interdot (capture and relaxation) process. Thus at first, with increasing current, carriers are relaxed fast from WL to gs through es. Then, gs have the main contribution as shown in the left inset of Fig. 3. With more current increment, the fast interdot relaxation becomes not the controlling factor as the system gets the saturation and a limit cycle was obtained in the system behavior as in the right inset of Fig. 3. The right inset show more carriers are still with WL while es has less carrier as it is an intermediate state (reservoir) and their carriers are relaxed to gs. This is with our recent conclusion that WL washes-out modes [16].

In Fig.4,the bifurcation diagram of photons vs. delay time reveals very complex patterns in the laser output. The laser develops into chaos states through a normal pulsing path, doubling phase, quasi-periodic, and finally a chaotic pulsing state. This is with the conclusion of Fig. 3. A limit cycle was obtained at a short delay time (high carrier density). The chaotic activity has been achieved for a long delay (less carrier density).Other than sporadic and quasi-periodic bifurcations in our system of frequency relaxation oscillations, which are common routes to chaos,wo= 0.001(blue curve), the sporadic path to anarchy is one of the well-observed roads. LFF is one of the unstable oscillations often found in QD lasers with optical feedback, known as the periodic chaos of saddle-node instability.Fig.4 also shows a sudden power dropout with a following gradual power recovery with wo= 0.02 (red curve) and wo= 0.05(green curve). From the perspective of realistic applications, the study of LFFs is very important because LFFs induces a lot of noise in the laser output power[36].

To display the phase of power recovery, the local limit of LFFs is averaged and plotted in Fig. 5 wherea stepwise power retrieval is evident. Relaxation oscillations are used in the power recovery versus each hop and the period of each step in the power recovery is typical of the round trip time of light in the external loop.

Fig.6 shows the bifurcation diagram vsthe feedback strength at bias current $\delta o = 0.3$. For small feedback strength ($Q< 0.27$) the laser shows a steady-state operation at the first cavity mode. The stable limit cycle loses at $Q = 0.3$ into a solution with periodically modulated photon number. For$Q > 0.31$,the bifurcation diagram shows two frequencies: the maxima and minima of the limit loop. The instability found in QD lasers, in this case, is quasi-periodic oscillation and a turbulent state.The laser output displays frequent pulsing with continuous pulse peak amplitude and interval as the feedback strength increases. The laser leads to a quasi-periodic path through erratic pulsing states for better feedback.The periodicity in chaotic behavior in QD laser is also detected in [27].

Fig. 7 shows thetime series of photon number, the carrier density of QD gs, es, and WL attractor, and ISI for three chosen values of feedback strength where Fig. 7 (a) and (b) are taken from the above figures. First, in Fig. 7 (a), the feedback strength is taken as Q=0.22where its current dependence is examined in Fig. 3. It gives periodic pulsations corresponding to a limit cycle shown in the second column of this figure. The second point chosen is Q=0.3,it is indicated earlier in the bifurcation diagram in Fig. 4. It is shown thatthis further increase in the feedback strength Q, the system undergoes a period-doubling route to chaos with windows of period two. The corresponding time series, attractors, and ISI are depicted in Fig. 7 (b), which is related to the two-folded limit cycles in the phase space projection shown in the same figures. InFig. 7 (c) the feedback strength is taken Q = 0.31 where the laser output is chaotic as seen in the broad spectrum of Fig. 7 (c) and the large chaotic attractor in the phase space projections. This leads to a time series that is susceptible to initial circumstances, such as spikes on top of a turbulent backdrop. The corresponding ISI histogram reveals that as its exponential tails suggest, theperiodic (chaotic) history causes the spikes in an irregular sequence. However, the ISI histogram reveals a complex structure with sharp peaks on top of this history, showing that the complex structure of unstable periodic orbits embedded in the chaotic attractor depends on Q.

II. Conclusion

In this work, a dimensionless model for QD laser dynamics, under external optical feedback, was presented where WL and QD states (GS and ES) were considered.The decrease of the capture rate value leads to a slowdown in the turn-on value in addition to a change in the value of the equilibrium point, and the reason is due to a slowdown in the speed at which the charge carriers move from the wet layer to the excited layer. Stepwise energy recovery is noticeable oscillations of relaxation that are used in the power recovery versus each hop and the period of each step in the power recovery is the property of the round trip time of light in the external loop.The system goes to periodic oscillations at a high injection current, low feedback rate, or delay time.This can be reasoned to fast interdot relaxation.

References

I. Fischer, Y. Liu, and P. Davis: Synchronization of chaotic semiconductor laser dynamics on subnanosecond time scales and its potential for chaos communication, Phys. Rev. A 62, 011801 (2000).

T. Heil, J. Mulet, I. Fischer, C. R. Mirasso, M. Peil, P. Colet, and W. Elsäßer: On/off phase shift keying for chaos-encrypted communication using external-cavity semiconductor lasers, IEEE J. Quantum Electron. 38, 1162–1170 (2002).

R. Vicente, C. R. Mirasso, and I. Fischer: Simultaneous bidirectional message transmission in a chaos-based communication scheme, Opt. Lett. 32, 403–405(2007).

A. Argyris, D. Syvridis, L. Larger, V. Annovazzi-Lodi, P. Colet, I. Fischer, J. García-Ojalvo, C. R. Mirasso, L. Pesquera, and K. A. Shore: Chaos-based communications at high bit rates using commercial fibre-optic links, Nature 438, 343–346 (2005).

E. A. Viktorov, P. Mandel, and G. Huyet: Long-cavity quantum dot laser, Opt. Lett. 32, 1268–1270 (2007).

W. Kinzel, A. Englert, and I. Kanter: On chaos synchronization and secure communication, Phil. Trans. R. Soc. A 368, 379–389 (2010).

M. Radziunas, A. Glitzky, U. Bandelow, M. Wolfrum, U. Troppenz, J. Kreissl, and W. Rehbein: Improving the Modulation Bandwidth in Semiconductor Lasers by Passive Feedback, IEEE J. Sel. Top. Quantum Electron. 13, 136–142 (2007).

V. Z. Tronciu: Excitability and coherence resonance of a DFB laser with passive dispersive reflector, Moldavian Journal of the Physical Sciences 7, 516 (2008).

S. Schikora, P. Hövel, H. J. Wünsche, E. Schöll, and F. Henneberger: Alloptical noninvasive control of unstable steady states in a semiconductor laser, Phys. Rev. Lett. 97, 213902 (2006).

V. Flunkert and E. Schöll: Suppressing noise-induced intensity pulsations in semiconductor lasers by means of time-delayed feedback, Phys. Rev. E 76, 066202 (2007).

T. Dahms, P. Hövel, and E. Schöll: Control of unstable steady states by extended time-delayed feedback, Phys. Rev. E 76, 056201 (2007).

T. Dahms, P. Hövel, and E. Schöll: Stabilizing continuous-wave output in semiconductor lasers by time-delayed feedback, Phys. Rev. E 78, 056213 (2008).

B. Fiedler, S. Yanchuk, V. Flunkert, P. Hövel, H. J. Wünsche, and E. Schöll: Delay stabilization of rotating waves near fold bifurcation and application to all-optical control of a semiconductor laser, Phys. Rev. E 77, 066207 (2008).

T. Dahms, V. Flunkert, F. Henneberger, P. Hövel, S. Schikora, E. Schöll, and H. J. Wünsche: Noninvasive optical control of complex semiconductor laser dynamics, Eur. Phys. J. ST 191, 71 (2010).

H. Al Husseini, S.F. Abdalah, K. Al Naimee, R. Meucci, F.T. Arecchi, " Exploring phase control in a quantum dot light-emitting diode,"Nanomaterials and Nanotechnology Volume 8: 1–7, 1-7, (2018).

Hussein B. Al Husseini, Kais A. Al Naimee, Amin H. Al-Khursan, and Ali. H. Khedir," External modes in quantum dot light emitting diode with filtered optical feedback." Journal of Applied Physics 119, 224301 (2016); DOI: 10.1063/1.4953651. View online: http://dx.doi.org/10.1063/1.4953651.

Hussein B. Al Husseini, and Hussein A. Al Rekabie, "Optical Injection Locking in Quantum Dot Light Emitting Diode."Journal of Applied Physics & Nanotechnology, Vol. 1 • Issue 1 • pp 1-5,(2018).

H. Al Rekabie, and H. Al Husseini, " Controlling of the Quantum Dot LED Dynamics with a Small Optical Feedback Strength "L&H Scientific Publishing, Journal of Applied Nonlinear Dynamics, 9, 57–70, 2020.

M. C. Soriano, J. García-Ojalvo, C. R. Mirasso, and I. Fischer: Complex photonics: Dynamics and applications of delay-coupled semiconductors lasers, Rev. Mod. Phys. 85, 421–470 (2013).

H. Su, L. Zhang, A. L. Gray, R. Wang, T. C. Newell, K. J. Malloy, and L. F. Lester: High external feedback resistance of laterally loss-coupled distributed feedback quantum dot semiconductor lasers, IEEE Photonics Technol. Lett. 15, 1504–1506 (2003).

G. Huyet, D. O'Brien, S. P. Hegarty, J. G. McInerney, A. V. Uskov, D. Bimberg, C. Ribbat, V. M. Ustinov, A. E. Zhukov, S. S. Mikhrin, A. R. Kovsh, J. K. White, K. Hinzer, and A. J. SpringThorpe: Quantum dot semiconductor lasers with optical feedback, Phys. stat. sol. (b) 201, 345–352 (2004).

D. O'Brien, S. P. Hegarty, G. Huyet, J. G. McInerney, T. Kettler, M. Lämmlin, D. Bimberg, V. Ustinov, A. E. Zhukov, S. S. Mikhrin, and A. R.

Kovsh: Feedback sensitivity of 1.3 μm InAs/GaAs quantum dot lasers, Electronics Letters 39, 1819–1820 (2003).

O. Carroll, I. O'Driscoll, S. P. Hegarty, G. Huyet, J. Houlihan, E. A. Viktorov, and P. Mandel: Feedback induced instabilities in a quantum dot semiconductor laser, Opt. Express 14, 10831–10837 (2006).

G. Huyet, D. O'Brien, S. P. Hegarty, J. G. McInerney, A. V. Uskov, D. Bimberg, C. Ribbat, V. M. Ustinov, A. E. Zhukov, S. S. Mikhrin, A. R. Kovsh, J. K. White, K. Hinzer, and A. J. SpringThorpe: Quantum dot semiconductor lasers with optical feedback, Phys. stat. sol. (b) 201, 345–352 (2004).

H. Al Husseini, and H. Al Rekabie, " Selecting Dynamics of the Quantum Dot Light Emitting Diode with a Small Optical Feedback Strength" ELSEVIER, Chaos, Solitons& Fractals, Nonlinear Science, and Nonequilibrium and Complex Phenomena, 118, 199–206, (2019).

K. Al Naimee, H. Al Husseini, S.F. Abdalah, A. Al Khursan, A.H. Khedir, R. Meucci, F.T. Arecchi, "Complex dynamics in Quantum Dot Light Emitting Diodes," Eur. Phys. J. D, 69: 257, 1-5, (2015).

B. Globisch, C. Otto, E. Schöll, and K. Lüdge: Influence of carrier lifetimes on the dynamical behavior of quantum-dot lasers subject to optical feedback, Phys. Rev. E 86, 046201 (2012).

V. Voignier.'stabilization of self-focusing instability in wide-aperture semiconductor laser',Phys. Rev. A, Volume:65, Issue, 5, Article Number: 053807(5 pages), (2002).

A. Uskov, Y. Boucher, J. L. Bihan, and J. McInerney, Appl. Phys. Lett. 11, 1499 (1998).

M. Sugawara, K. Mukai, and H. Shoji, Appl. Phys. Lett. 71, 2791 (1997).

H. Al Husseini, A. Al Khursan, K. Al Naimee, S.F. Abdalah, A.H. Khedir, R. Meucci, F.T. Arecchi, "Modulation Response, Mixed mode oscillations and chaotic spiking in Quantum Dot Light Emitting Diodes,"ELSEVIER, Chaos, Solitons& Fractals, Nonlinear Science, and Nonequilibrium and Complex Phenomena, 78, 229–237, (2015).

Hussein B. Al Husseini, Kais A. Al Naimee, Ali. H. Khedir, and Amin H. Al-Khursan," Dynamics of Quantum Dot Light Emitting Diode with Filtered Optical Feedback." Nanomaterials and Nanotechnology journal. Volume 6: 1–9, 2016.

Amin H. Al-Khursan, "Intensity Noise Characteristics in Quantum-Dot Lasers: Four-Level Rate Equations Analysis", J. Luminescence, 113, 129-136 (2005).

M. Gioannini, Amin H. Al-Khursan, G. A. P. Thé, and I. Montrosset, "Simulation of Quantum Dot Lasers with Week External Optical

Feedback", Dynamics Days Conference, Delft-Netherlands, 25-29 August, 2008. P. 167.

J.Ohtsubo, Semiconductor lasers stability, instability and chaos, Second Edition, Springer-Verlag Berlin (2008).

T.Sano,Antimode dynamics and chaotic itinerancy in the coherence collapse of semiconductor lasers with optical feedback. Phys Rev A 50:2719– 2726, (1994).

Made in United States
Orlando, FL
04 November 2026

90200293R00020